Easy-Step

Wiring

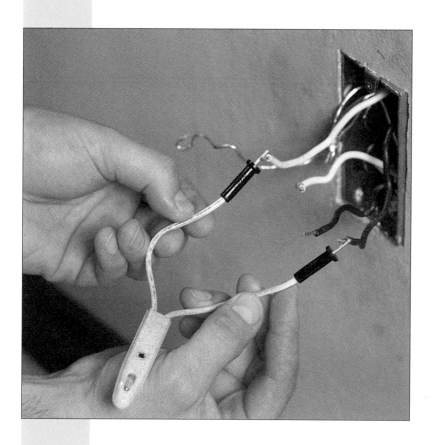

Contents

Introduction 4

Getting Started 7
1. Understand home wiring systems
2. Obtain basic tools
3. Select materials
4. Learn basic techniques
5. Practice safety
6. Use testers

Extending a Circuit 17
1. Plan the installation
2. Cut out box openings: wall
3. Cut out box openings: ceiling
4. Run cable
5. Install new box in wall
6. Install box in ceiling
7. Connect new receptacle

Wiring Switches and Fixtures 27
1. Connect single-pole switch
2. Install light fixture
3. Wire 3-way switch
4. Wire 4-way switch
5. Install ceiling fan
6. Install track lighting
7. Add recessed downlight
8. Install outdoor receptacle

Installing Surface Wiring   **41**
   1 Plan layout
   2 Install starter box
   3 Install channels, wires
   4 Wire receptacles, switches
   5 Wire fixtures

Upgrading Fixtures   **47**
   1 Replace 120-volt receptacle
   2 Replace 240-volt receptacle
   3 Install GFCI receptacle
   4 Replace switch
   5 Install dimmer

Making Repairs   **53**
   1 Replace socket
   2 Rewire lamp
   3 Replace lamp plug
   4 Add cord switch
   5 Repair doorbell

Adding Improvements   **59**
   1 Wire smoke detector
   2 Wire CO detector system
   3 Add motion detector
   4 Add surge suppressors
   5 Add motor watchman

U.S./Metric Measure Conversions   **64**

Introduction

Sooner or later, every home requires some type of electrical repair or improvement. It may be updating a light fixture in the kitchen, upgrading a bathroom outlet to be a ground fault circuit interrupter (GFCI) receptacle, or installing a ceiling fan in the bedroom. Many homeowners hesitate to make such improvements because they find it too costly to hire a licensed electrician—if, indeed, they can find one willing to take on small jobs. Others are leery of doing the work themselves because it seems too technical or dangerous.

Certainly, electricity deserves caution and respect. But it also is easy to understand. It functions according to precise mathematical principles and is completely logical. If you are reasonably handy, you can safely complete dozens of electrical projects—and enjoy the satisfaction of doing the work yourself—once you have learned a few basic techniques.

Wiring is your one-stop guide to learning these techniques. In simple steps, it leads you through typical residential repairs and upgrades. For best results, read through the first chapter to learn the fundamental principles and techniques of electrical work. In addition, study the section on connecting a new receptacle (pages 24 and 25) to see how these basics of home wiring apply to a specific project.

INTRODUCTION

For simplicity's sake, *Wiring* assumes that all wiring is done with nonmetallic-sheathed cable, commonly called Romex,® but flexible metal cable or insulated wire and conduit follow the same principles—except that a separate ground wire is not required with conduit. Metal boxes are shown to illustrate how they must be grounded; plastic boxes can also be used, but don't require grounding.

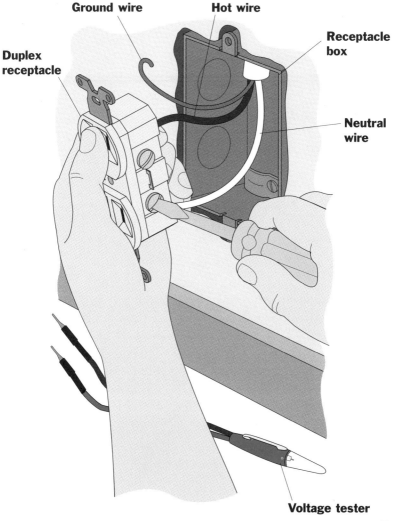

Getting Started

GETTING STARTED

1 Understand home wiring systems

Two hot wires and one neutral wire carry electricity to the service panel, where it is routed to the home's various circuits. Each circuit has a breaker or fuse that cuts off power when the circuit fails. Most homes have 120-volt circuits for lights, receptacles, and small appliances, and 240-volt circuits for heavy appliances. A 120-volt circuit has a hot wire (black or red), a neutral wire (white), and a ground wire (green or bare copper) that carries electricity safely to the earth when the normal flow is interrupted.

TIP: In some homes, metal conduit or armored cable is used for grounding instead of a separate ground wire.

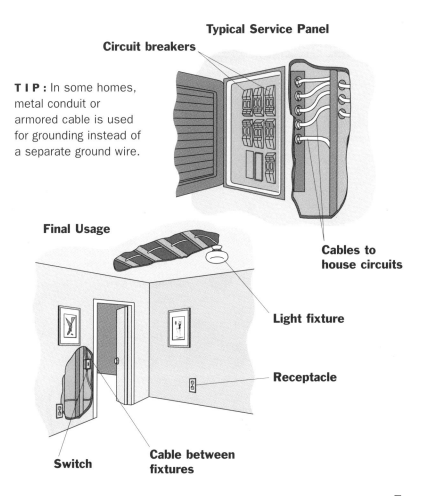

GETTING STARTED

2 Obtain basic tools

Buy or rent the tools shown below. Some are carpentry tools used to cut and drill openings or to fasten electrical components. Others are more specialized, including needle nose pliers for making wire loops and reaching into tight places, a multipurpose tool for cutting and stripping wire, linesman's pliers for cutting cable, and a cable stripper for slitting cable sheathing. The fish tape, used for pulling cable through walls or wire through conduit, can be rented. To use the voltage and continuity testers, see pages 14 and 15.

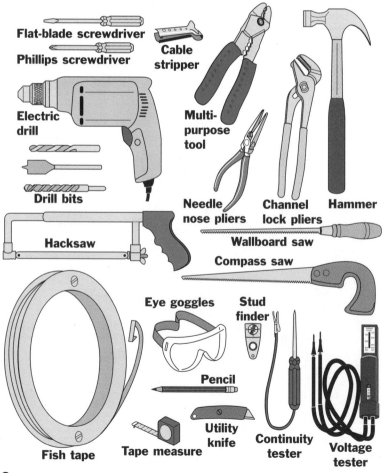

GETTING STARTED

Select materials

Most jobs require electrical boxes, sheathed cable or wire and conduit, and appropriate connectors. Boxes can be metal or plastic. "Old-work" boxes attach to wall or ceiling materials with clamps and plaster ears. "New-work" boxes attach directly to bare framing. Be sure the boxes have internal clamps for the type of cable you use. Select the deepest box that will fit in the space. Choose the right wire for the job: No. 14 for 15-amp (normal household) circuits; No. 12 for 20-amp circuits (small appliances).

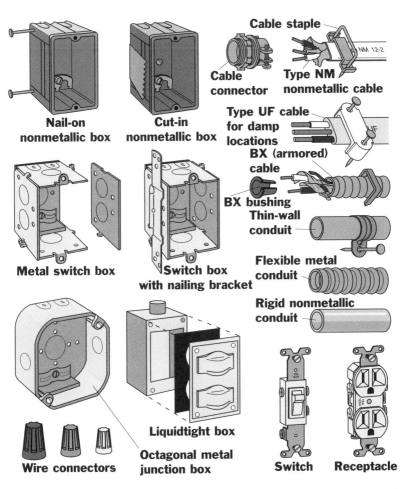

GETTING STARTED

Learn basic techniques

To strip away cable sheathing, also called jacket, use a cable stripper. Slide the stripper about 8 inches onto the cable. Holding the cable, squeeze the stripper shut and pull the cable back. Peel the slit sheathing away from the wire and remove it. Check that the wire insulation is not damaged. To strip wire, use a multipurpose tool or wire stripper. Lay the wire in the correct groove for its size, squeeze the handles together, and pull off the correct amount of insulation as indicated by the device's strip gauge.

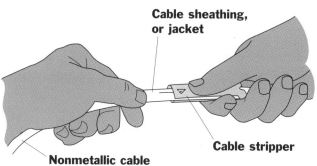

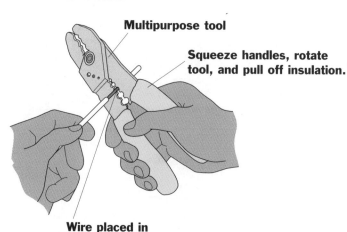

GETTING STARTED

To connect a wire to a screw terminal, strip ¾ inch of insulation from the wire end. Using needle nose pliers, form the stripped end of the wire into a circle. Hook the loop clockwise around the screw terminal and tighten. The screw head should cover the entire loop. Cut off any excess. To join wires with wire connectors, strip away ½ inch of insulation from both wires, hold them together, and push them into the connector. Twist the connector clockwise until tight. Pull on each wire to test the connection.

Joining Wires to Screw Terminals

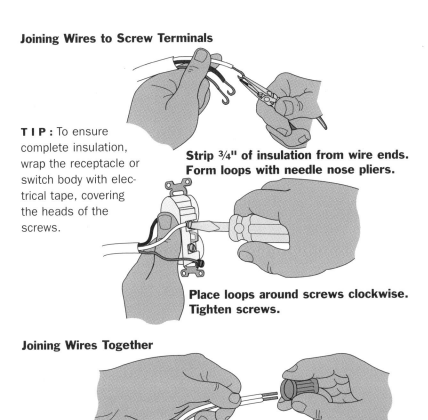

TIP: To ensure complete insulation, wrap the receptacle or switch body with electrical tape, covering the heads of the screws.

Strip ¾" of insulation from wire ends. Form loops with needle nose pliers.

Place loops around screws clockwise. Tighten screws.

Joining Wires Together

Strip ½" of insulation from wire ends. Twist a wire connector of the proper size clockwise onto the ends.

GETTING STARTED

5 Practice safety

Always turn off the power at the service panel before working on the electrical system. Switch off the panel's main disconnect or turn off the breaker for the specific circuit. To find the circuit, plug in a radio and turn up the volume. Loosen fuses or switch off circuit breakers one at a time until the radio stops playing. Remove the fuse or tape the breaker in the *off* position.

Switch off the circuit. Post a note to warn others. Lock or tape the panel door shut.

On damp floors, work standing on a dry rubber mat or dry boards.

Lock the main breaker panel, or post a sign that warns others that you're working on the system. Stand on dry boards or a rubber mat if the floor is wet or damp. Never touch metal objects such as pipes or appliances while working with electricity. Always use tools with plastic or rubber-coated handles and wear gloves when possible. Wear safety goggles to protect your eyes while sawing, hammering, or drilling.

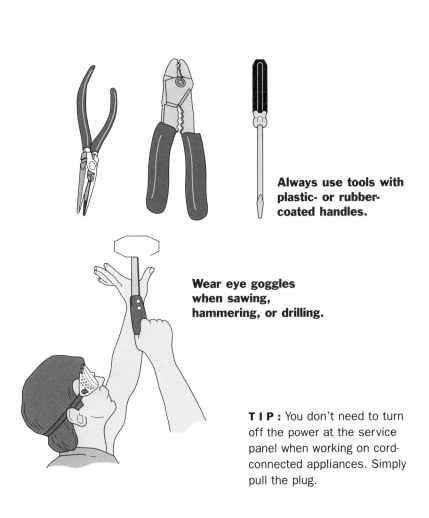

Always use tools with plastic- or rubber-coated handles.

Wear eye goggles when sawing, hammering, or drilling.

TIP: You don't need to turn off the power at the service panel when working on cord-connected appliances. Simply pull the plug.

GETTING STARTED

6 Use testers

Use a voltage tester to make sure the circuit is dead before you begin work, and to check for power and proper grounding when you're done. Before turning off any power, test the tester itself on a live circuit to be sure it works. The least expensive tester has a small neon lamp with two short test leads. Buy several. Use as shown below to check for power and proper grounding. Check all combinations of wires. The tester lights up when one lead touches a hot wire and the other lead touches a neutral wire or a ground.

Using a voltage tester to test for power at a receptacle

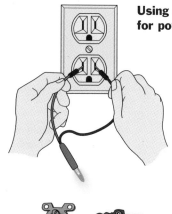

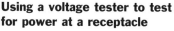

For grounded receptacles, left-hand probe should be in U-shaped grounding hole or grounded screw.

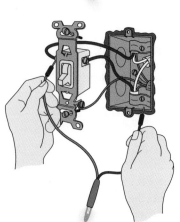

Finding a hot wire

14

Use a battery-powered continuity tester to test disconnected appliances, switches, and wiring for complete circuits. One common type looks like a pocket flashlight with a probe on one end and a long test lead on the other. A volt-ohmmeter, or multitester, can also be used. Turn off the power to the electrical device before testing. The lamp lights up or the needle responds when a circuit is completed between the probe and the test lead.

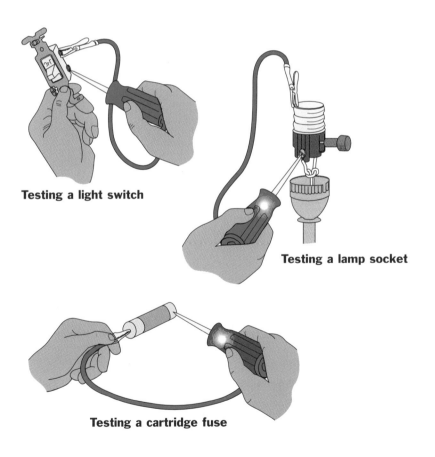

Testing a light switch

Testing a lamp socket

Testing a cartridge fuse

Extending a Circuit

EXTENDING A CIRCUIT

1 Plan the installation

When installing a new receptacle, locate it on an interior wall accessible from a basement or attic. Feed new cable to it from a nearby existing receptacle. First deaden the circuit, open the existing receptacle's box, and examine its interior. If it has an unused knockout, route the cable from it to the new box. Run the cable vertically through the wall cavity, then horizontally through or along joists, and then up or down to the new box. You also can route cable behind high baseboards.

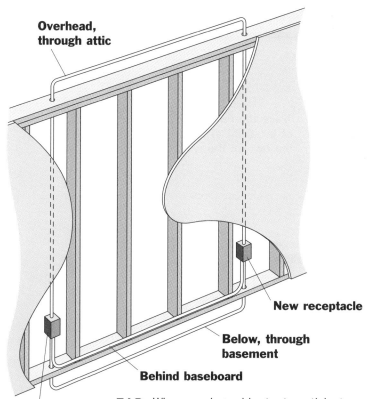

Overhead, through attic

New receptacle

Below, through basement

Behind baseboard

Existing receptacle

TIP: When running cable, try to anticipate obstructions such as insulation, air ducts, blocking, and plumbing. Ideally, expect 14½ inches of hollow space between studs.

17

EXTENDING A CIRCUIT

2 Cut out box openings: wall

Locate studs with a stud finder or a piece of stiff wire inserted through a small hole. Center a new box between studs, at the same height as other boxes in the room. Adjust the box for plumb and trace its outline. Score the outline with a utility knife, guided by a straightedge. Position the point of a wallboard saw on one corner of the outline and gently hammer the handle with your palm until the blade penetrates the wallboard. Saw along the line. Poke through and saw each line until the cutout is complete.

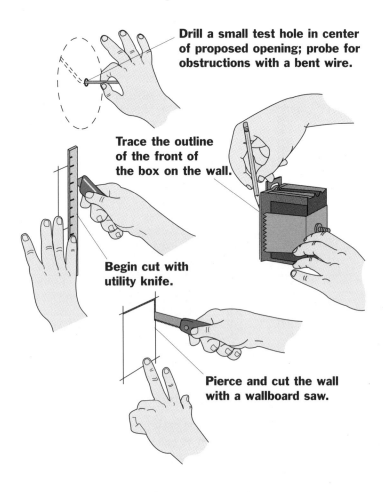

Drill a small test hole in center of proposed opening; probe for obstructions with a bent wire.

Trace the outline of the front of the box on the wall.

Begin cut with utility knife.

Pierce and cut the wall with a wallboard saw.

EXTENDING A CIRCUIT

3. Cut out box openings: ceiling

To center a box in the ceiling, measure diagonals from opposite corners and make a small hole where they intersect. Insert a wire through the hole to locate ceiling joists. Position the new box between joists, as close as possible to the hole, and trace its outline on the ceiling. Use a wallboard saw to cut the opening. (If access is easier, make the cutout from above. First mark the box location, drill a small hole in the center, and push a wire through it. From above, locate the wire, trace the box outline, and cut.)

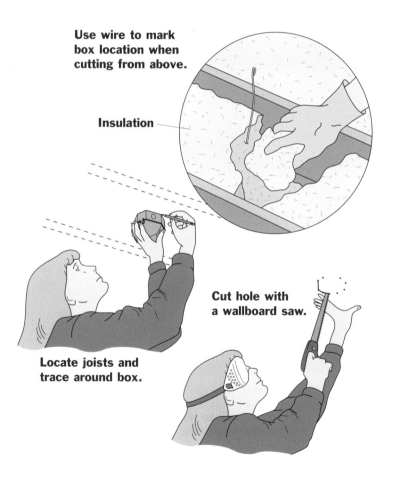

Use wire to mark box location when cutting from above.

Insulation

Locate joists and trace around box.

Cut hole with a wallboard saw.

EXTENDING A CIRCUIT

 Run cable

Use cable sized for the circuit breaker (see page 9). Before drilling holes through wall plates, drill small pilot holes at an angle from below and push short wires through them for guides. From above, drill ¾-inch holes through the wall plates and the centers of any joists in the way. Turn off power to the source receptacle and remove a knockout from the box. Using fish tape, pull cable from the new box location, into the attic, and to the source box. Cable ends should extend 6 inches beyond boxes and 12 inches beyond cutouts without boxes.

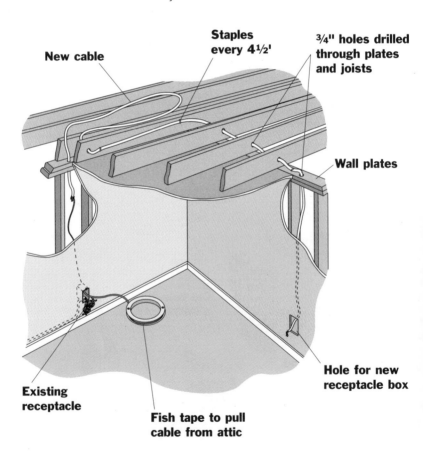

EXTENDING A CIRCUIT

When running cable from below, the greatest difficulty is finding the center of the wall cavity. Pipes or ducts are helpful guides to take measurements from. Or, drill a small angled hole toward the basement from the bottom of the baseboard and push a short wire into it as a locator. Drill holes and fish the cable as described on page 20.

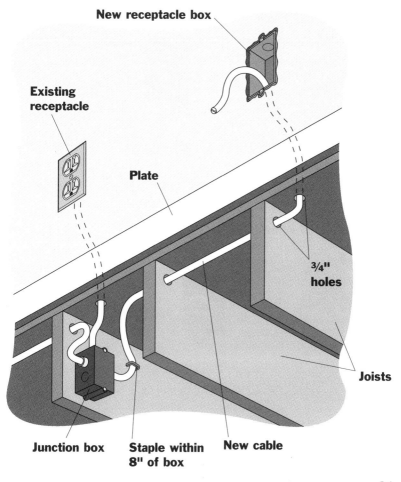

EXTENDING A CIRCUIT

5 Install new box in wall

Choose an old-work, or "cut-in," box—one designed to fasten to wall material, not framing. Remove a top or bottom knockout, depending on which way the cable approaches the box. Feed the cable into the box, pulling it 12 inches beyond the front. Slip the box into its opening and secure it in place. Tighten the cable clamp.

Pull the cable through a knockout hole in the box.

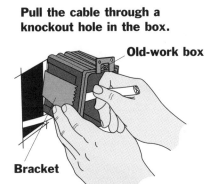

Old-work box

Bracket

Push box into place.

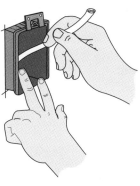

Tighten bracket screw.

Tighten cable clamp.

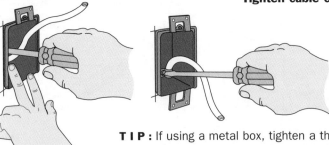

TIP: If using a metal box, tighten a threaded connector onto the cable before pulling it into the box; then slip the locknut over the end of the cable and tighten it onto the connector threads from inside the box.

EXTENDING A CIRCUIT

6 Install box in ceiling

A ceiling box can be installed like a wall box unless it is intended to support a heavy fixture. Large light fixtures or ceiling fans require boxes specifically certified for that use. Such a box must be solidly attached between two ceiling joists. If you have access from above, nail a brace between the joists and mount the box to it, flush with the ceiling. If not, reach up through the hole and tighten an expandable bar hanger in place and install the box included with the hanger.

Check for positioning between joists.

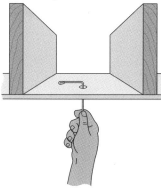

Cut hole for heavy-duty box, run cable.

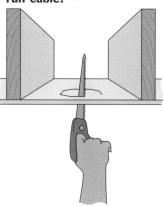

Install expandable bar hanger.

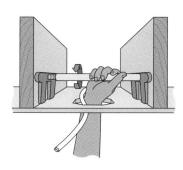

Pull cable through box; secure box to hanger.

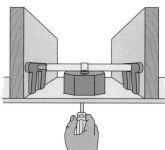

EXTENDING A CIRCUIT

7 Connect new receptacle

Review wire-handling techniques on pages 10 and 11. Turn off the circuit. After fishing the cable, remove the sheathing from both ends. In the new box, trim each wire to 6 inches and strip insulation from the end. Connect the black wire to the brass-colored screw of the receptacle, the white wire to the silver-colored screw, and the green (or bare) wire to the green screw. If using a metal box, pigtail the green (or bare) wire to a wire going to the receptacle's green screw and to the tapped hole in the box.

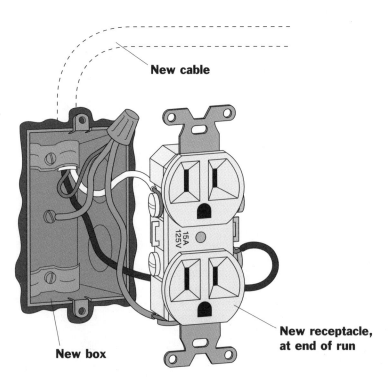

EXTENDING A CIRCUIT

Test the wires in the source box to be sure the circuit is dead. Remove the receptacle. Restrip the ends of the wires, if necessary, leaving them 6 inches long. Using wire connectors, connect the new black wire and a 6-inch pigtail to the ends of the other black wires. Do the same for the white and green (or bare) wires. Then attach the three pigtails to the receptacle. Push the receptacle and wires into the box, keeping bare ground wires away from brass-colored screws. Secure the receptacle to the box and attach the cover plate.

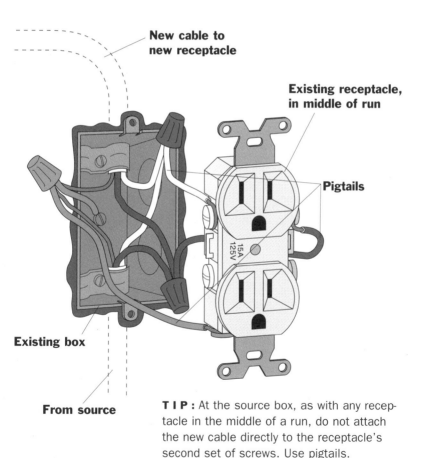

New cable to new receptacle

Existing receptacle, in middle of run

Pigtails

Existing box

From source

TIP: At the source box, as with any receptacle in the middle of a run, do not attach the new cable directly to the receptacle's second set of screws. Use pigtails.

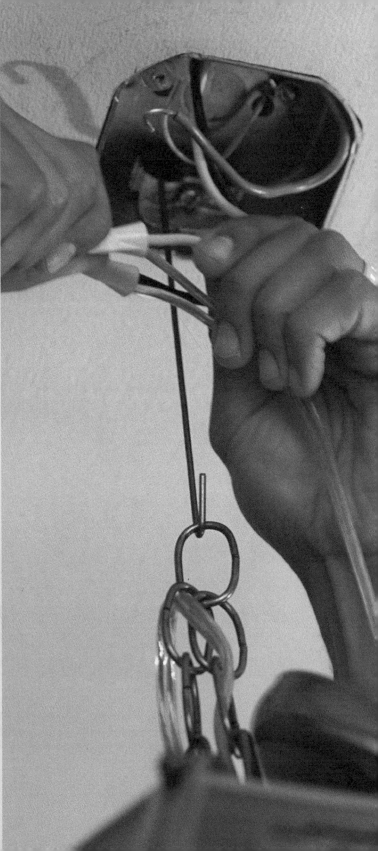

Wiring Switches and Fixtures

WIRING SWITCHES AND FIXTURES

Connect single-pole switch

A single-pole switch, also called a 2-way switch, operates a light or other fixture from only one location. It has two terminal screws. Before wiring, be sure the circuit is off. If power comes through the switch to the fixture, splice the white wires with a wire connector, connect the black wires to the screws (it doesn't matter which goes to which), and attach the ground. If power comes to the fixture first, both wires that go to the switch are considered hot, even if one is white. Attach either wire to either screw. Connect grounds.

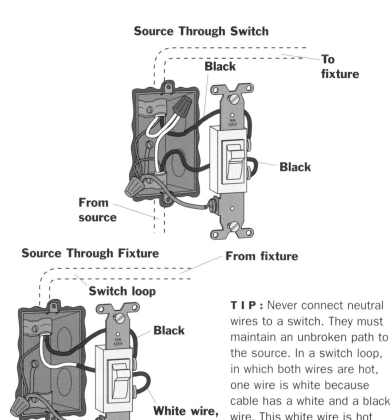

TIP: Never connect neutral wires to a switch. They must maintain an unbroken path to the source. In a switch loop, in which both wires are hot, one wire is white because cable has a white and a black wire. This white wire is hot and is attached to the switch.

WIRING SWITCHES AND FIXTURES

2 Install light fixture

If wires come from the power source through the switch, connect the white wire to the fixture's silver-colored screw or white pigtail, and the black wire to the brass-colored screw or black pigtail. If wires come to the fixture box from the source and then loop to a switch, disconnect the black wire from the source and connect it to the switch loop's white wire. Connect the white wire from the source to the fixture's silver-colored screw or white pigtail. Connect the switch loop's black wire to the fixture's brass-colored screw or black pigtail.

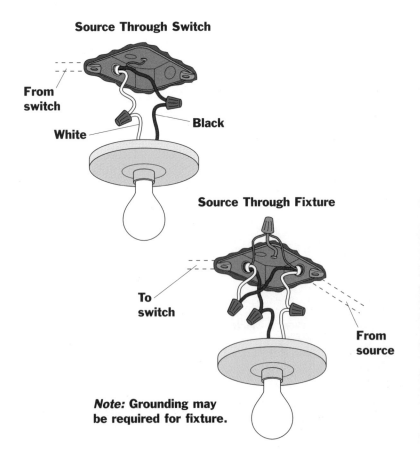

Source Through Switch

From switch
White
Black

Source Through Fixture

To switch

From source

Note: Grounding may be required for fixture.

28

WIRING SWITCHES AND FIXTURES

Wire 3-way switch

Used in pairs, 3-way switches operate fixtures from two locations. A 3-way switch has three screw terminals. One is a bronze-colored screw, labeled *Common,* for attaching the hot wire from the source or from the fixture, depending on which switch you are wiring. The other two terminals are brass-colored screws for attaching the two "traveler" wires that run between the two switches. These wires are considered hot. The neutral white must not be connected to either switch. Five arrangements are shown below and on pages 30 and 31.

3-Way, Source Through Central Fixture Box

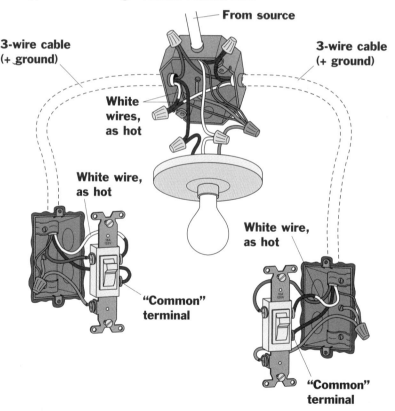

WIRING SWITCHES AND FIXTURES

Wire 3-way switch (con't.)

The three-wire switching arrangement shown at the top resembles the arrangement on page 29, but the power source comes to one of the switches first. In the second diagram, the power comes to the fixture first, with both switches beyond it.

3-Way, Source Through Switch Box (Central Fixture)

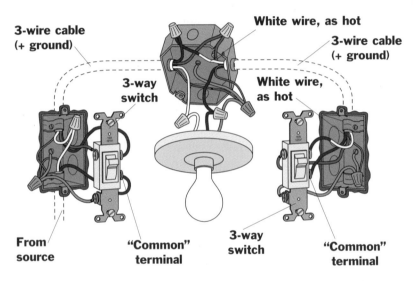

3-Way, Source Through End-Fixture Box

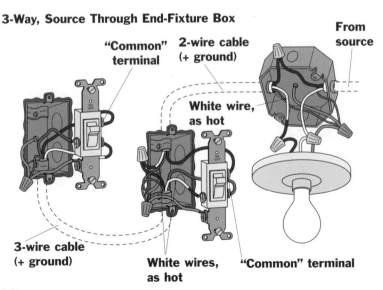

The top diagram below shows both switches wired on one side of the fixture. The power source comes to the end switch first. The lower diagram shows a similar arrangement, but there is a receptacle beyond the fixture that must always be hot and should not be affected by the switches.

3-Way, Source Through Switch Box (End Fixture)

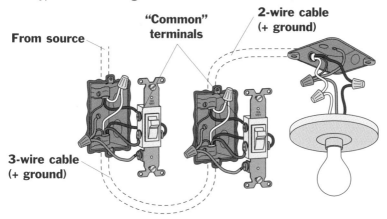

3-Way, Source Through Switch Box; Receptacle Beyond

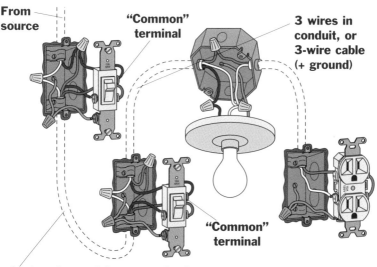

WIRING SWITCHES AND FIXTURES

Wire 4-way switch

To operate a fixture from three locations, install a 4-way switch between two 3-way switches. Wire as shown, connecting the black wire of the source box to the closest 3-way's "common" screw. Connect switches so that wires attached to this 3-way's brass-colored screws are attached to the 4-way's lower or upper pair of brass-colored screws. Wire the remaining pair to the second 3-way's brass-colored screws. Connect the black wire so the bronze-colored screw is wired to the fixture's "common" screw or black wire. Make neutral wire and ground connections.

4-Way, Source Through Fixture Box

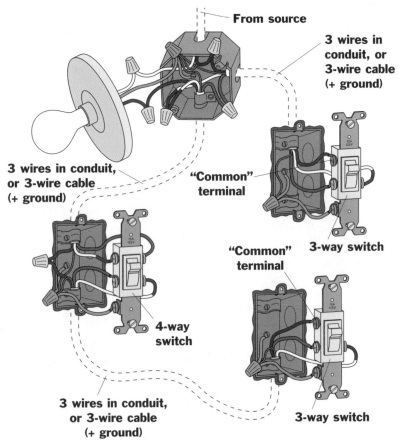

WIRING SWITCHES AND FIXTURES

5 Install ceiling fan

Use a ceiling box rated for fan support. See pages 19 and 23 for how to cut the opening and install the box. Feed power to the box in the same manner as for a light fixture (see page 28). Attach the fan hanger (or support bracket) to the box. Slip the stem and ball attached to the fan assembly into the hanger. Bring the fan's wires up through the stem into the ceiling box and connect them to the house wiring as shown. Assemble the fan, following manufacturer's instructions.

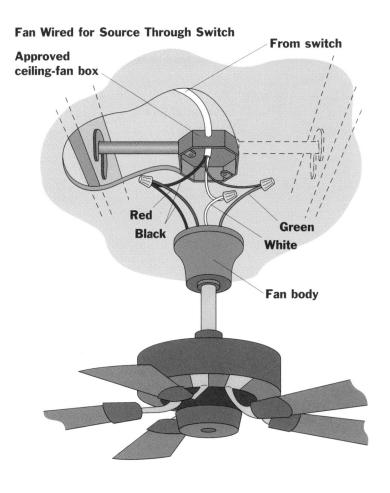

Fan Wired for Source Through Switch
Approved ceiling-fan box
From switch
Red
Black
Green
White
Fan body

WIRING SWITCHES AND FIXTURES

6 Install track lighting

Track-light installation differs slightly from manufacturer to manufacturer, so be sure all the system's parts are the same brand and follow manufacturer's instructions. You can convert a ceiling light to track lighting by using the same outlet box. If an existing box is not available where you want to install track lighting, see pages 19 and 23. Choose a midrun or end-run wire-in connector as the situation requires. With the old fixture removed, or the new wiring in place, wire the connector as shown and attach it to the ceiling box.

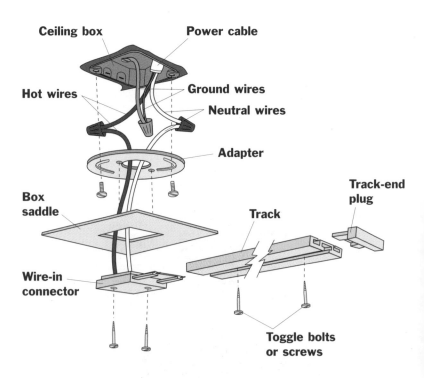

WIRING SWITCHES AND FIXTURES

Cut track to length with a fine-toothed hacksaw. Measure from the side of the adapter to the nearest wall. Mark this dimension on the ceiling in several places so the track will be parallel to the wall. Hold the track on the ceiling. Mark and drill holes for screws (if they will hit ceiling joists) or toggle bolts. Plug the track to the wire-in connector and fasten it to the ceiling. Use couplings to extend the track. Close the outer ends with end plugs. Install and position the lights. Most have a thumb switch for inserting them in the track and require a twisting motion to lock them in place.

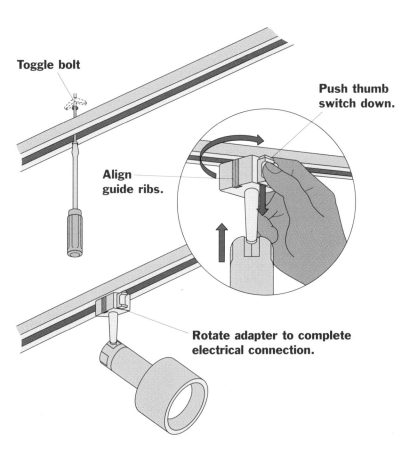

WIRING SWITCHES AND FIXTURES

7 Add recessed downlight

Recessed downlights that are designed for installation in ceilings without overhead access come with sheet-metal brackets, called plaster frames, and/or ceiling clips. Units labeled *T.C.* or *Non I.C. Housing* must be located at least 3 inches from insulation. Only units labeled *I.C. Housing* can contact insulation. Both types require cable or wire rated 90° C. Make the ceiling cutout (see page 19) using the manufacturer's template. Fish cable from a switch, up the wall, and between the joists to the cutout. Pull 2 feet of cable through the opening.

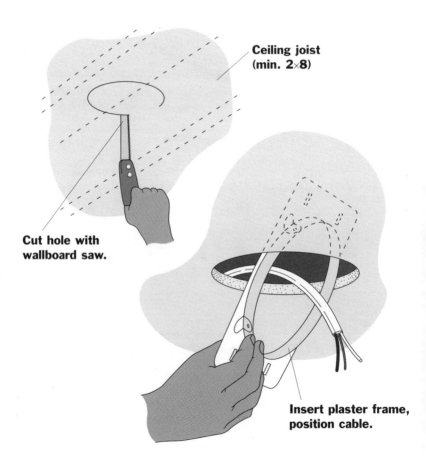

Ceiling joist (min. 2×8)

Cut hole with wallboard saw.

Insert plaster frame, position cable.

36

WIRING SWITCHES AND FIXTURES

Methods of securing the housing in the cutout vary. Some units have a plaster frame (shown on page 36) that can be tipped into the cutout and fastened with clips. Complete the wiring, push the housing into place, and fasten it to the plaster frame. Other types have clips that grip the back of the ceiling material when the unit, already wired, is pushed into the cutout. Once the wired housing is in place, attach the trim and install the lamp.

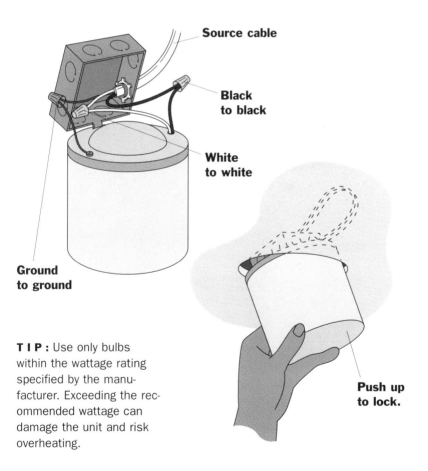

T I P : Use only bulbs within the wattage rating specified by the manufacturer. Exceeding the recommended wattage can damage the unit and risk overheating.

WIRING SWITCHES AND FIXTURES

Install outdoor receptacle

Install an outdoor receptacle in the same stud bay as the interior source receptacle, but not directly behind it. Use a window or door as a reference point for measuring the location of the cutout. To ease installation, position the cutout 3 inches from a stud. Hold a box in place and trace its outline onto the exterior wall. Drill holes in the corners and cut the opening with a keyhole saw. Turn off circuit. Open a knockout in the source box. Run cable between the boxes, leaving 8 inches at the source and 12 inches at the exterior.

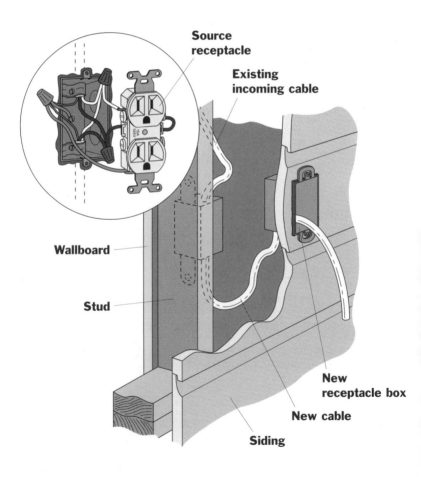

WIRING SWITCHES AND FIXTURES

Remove a knockout on the new box. Slip the box over the cable and push it into the opening. Clamp the cable in place, leaving about 8 inches excess. Strip away the cable sheathing and screw the box to the siding. You must use a ground fault circuit interrupter (GFCI) receptacle outdoors (follow manufacturer's instructions or see page 49). Install the gasket and liquidtight cover. Turn off the power. Inside, disconnect the receptacle. Strip away the cable sheathing. Pigtail to the cable with 6-inch pieces of scrap wire. Reinstall and cover.

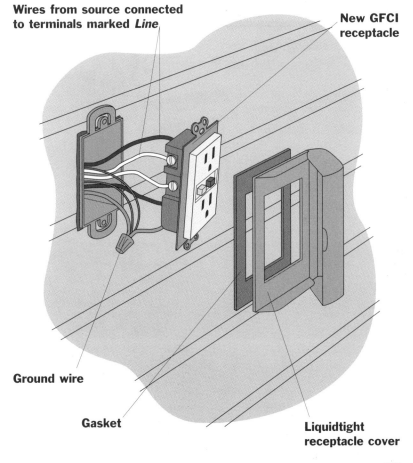

Installing Surface Wiring

INSTALLING SURFACE WIRING

1 Plan layout

Surface wiring lets you extend a circuit without breaking into a wall or ceiling. Begin by selecting a source box. Count the number of outlets on its circuit, including the additions. A 15-amp circuit handles 10 outlets; a 20-amp circuit handles 13. If the circuit has sufficient capacity, decide where to locate the new receptacles, switches, and fixtures. Plan a short, unobtrusive run from the source box to the new devices. Typical runs drop from the source box, run along the baseboard, and rise to the new devices.

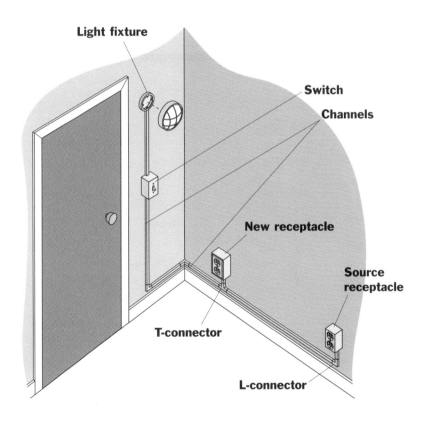

TIP: Baseboard raceway systems disguised as molding have multiple channels for electrical wiring and communications cables—ideal for home offices. Boxes combine receptacles and jacks.

INSTALLING SURFACE WIRING

2 Install starter box

Before you begin, check that all the system's parts are from the same manufacturer. Remove the cover plate of the source box. Remove the switch or receptacle from the box, but do not disconnect its wires. Place the starter-box base plate over the existing box and fasten it with the anchors provided in the manufacturer's accessory kit. An outlet-box base plate can be used as a starter-box base plate by removing the rectangular center knockout. Fasten it to the source box.

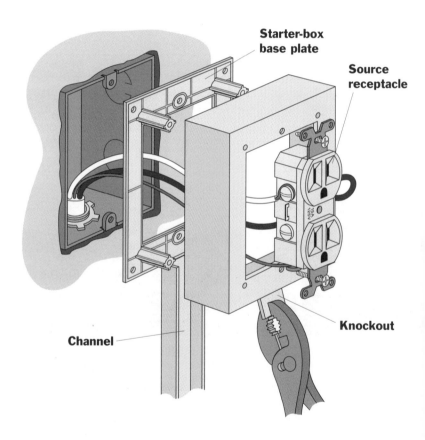

INSTALLING SURFACE WIRING

Install channels, wires

Beginning at the source box, hold the channels in place and mark them for cutting. Trim them to size with a hacksaw. Drill holes in the channel every 18 inches and 1 inch from ends. Level or plumb the channel and mark the locations of holes. Secure with wallboard anchors and screws. Break off the edges at T-connections to clear the channel between horizontal and vertical base pieces. Lay wire, using the same gauge as the source circuit. Allow an additional 12 inches for each device that will be wired. Hold the wire in place with snap-on clips.

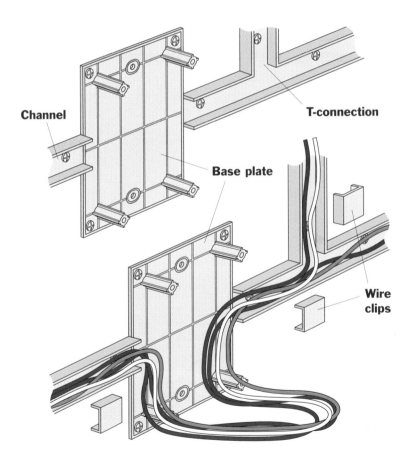

INSTALLING SURFACE WIRING

Wire receptacles, switches

Turn off circuit. Wire the source-box receptacle using 6-inch pieces of wire to pigtail like-colored wires together. Wire new receptacles (see pages 24 and 25), attach extension boxes to the base plates, and fasten receptacles in place. When installing a switch, strip and cut the black and green wires at the switch, but do not cut the white wire; it runs through the switch box to the fixture. Connect the black wires to the switch's two brass-colored screws. Attach a grounding pigtail to the switch and to the two ground wires.

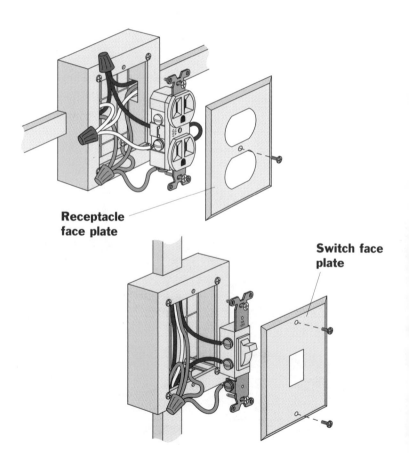

Receptacle face plate

Switch face plate

INSTALLING SURFACE WIRING

5 Wire fixtures

Remove the knockout where the fixture-box cover fits over the channel and attach the cover to the base plate with screws. Pull the wires through.

Strip ½ inch from the wires and pigtail them, the black wire to the fixture's black wire, the white wire to its white wire, and the green wire to its green wire. Then, following manufacturer's directions, mount the fixture on the fixture-box cover. Install the correct lamps and attach the globe or diffuser.

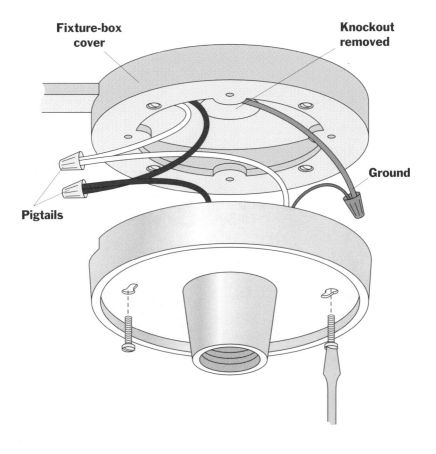

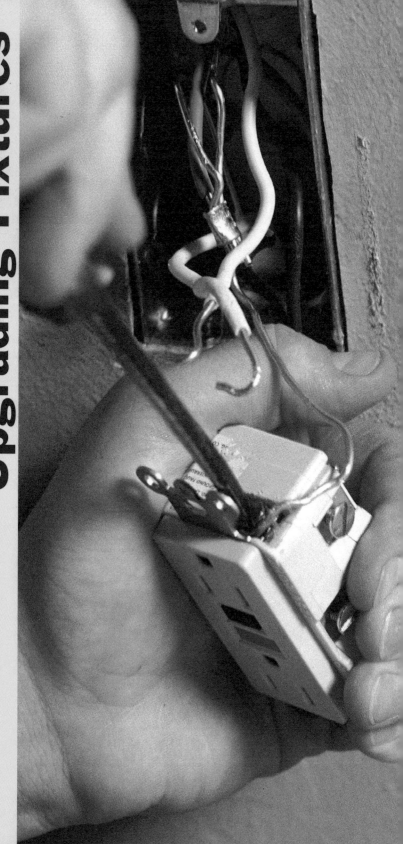

Upgrading Fixtures

UPGRADING FIXTURES

1 Replace 120-volt receptacle

Turn off circuit. If receptacle is at the end of a run, disconnect the wires. Restrip and connect them to the new receptacle. If the box is metal, add two grounding pigtails. For a receptacle in the middle of a run, look for a small link connecting the two brass-colored screws. If present, connect pigtailed wires, including the existing ground wire, to the new receptacle as shown. If link is removed, you have a "split-wired" receptacle, in which one outlet is controlled by a switch. Replace wire for wire.

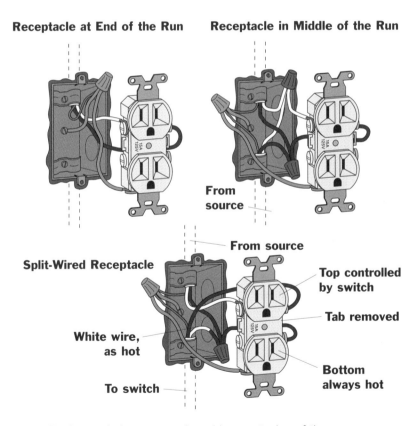

TIP: Replace existing receptacles with receptacles of the same rating. Most residential receptacles are rated 15 amps—even for 20-amp circuits (an exception allowed for residences).

UPGRADING FIXTURES

2 Replace 240-volt receptacle

Some 240-volt receptacles look like 120-volt receptacles, but have a horizontal slot and two brass-colored terminal screws, not one brass-colored and one silver-colored. Three-wire, 240-volt appliance receptacles have three slots of various configurations, depending on the rated amperage. Buy a new receptacle with the same configuration. To replace a receptacle, deaden the circuit and replace wire for wire. Attach a grounding pigtail to the box, unless conduit grounds the box. Tighten screws firmly, secure the receptacle in the box, and cover.

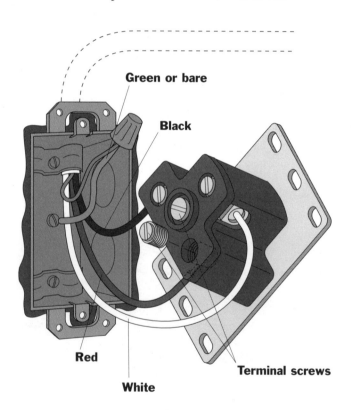

Green or bare

Black

Red

White

Terminal screws

UPGRADING FIXTURES

3 Install GFCI receptacle

Ground fault circuit interrupter (GFCI) receptacles cut off electricity when there is danger of shock. Use them where a person would be in contact with the earth or grounded water pipes—outdoors or in a kitchen, bathroom, garage, or unfinished basement. If the receptacle is at the end of a run or is connected with pigtails, simply connect the hot (black) and neutral (white) wires to the GFCI terminals marked *Line*. If you want the GFCI to protect all receptacles "downstream" from it, connect the outgoing (downstream) wires to the "load" terminals.

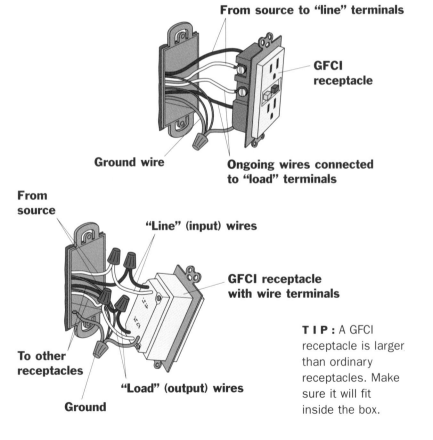

TIP: A GFCI receptacle is larger than ordinary receptacles. Make sure it will fit inside the box.

49

UPGRADING FIXTURES

Replace switch

To replace a 2-way switch, disconnect the wires, then trim and restrip them. Ground the new switch and connect either wire to either screw.

A 3-way switch is always paired with another 3-way switch and has three wires connected to it (besides the ground). Before removing the old switch, mark the wire attached to the bronze-colored terminal (labeled *Common*) and be sure to connect it to the "common" terminal of the new switch. Brass-colored terminals are interchangeable for the remaining two wires. See also pages 29 to 32.

2-Way Switch, with Source Through Switch

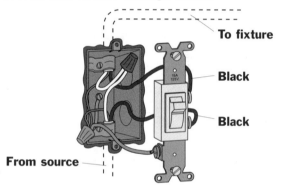

3-Way Switch

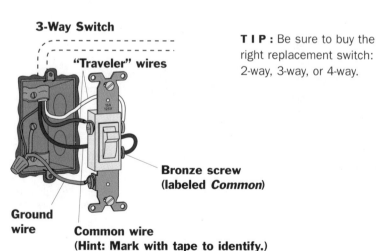

TIP: Be sure to buy the right replacement switch: 2-way, 3-way, or 4-way.

UPGRADING FIXTURES

5 Install dimmer

Buy a 2-way or 3-way dimmer, depending on the switch you are replacing. Remove the old switch. Trim and strip wires about ½ inch. If the dimmer has wires, strip them ½ inch as well. With a 2-way dimmer, either house wire can go to either of the two terminals. When replacing a 3-way switch, first use tape to tag the wire originally connected to the bronze-colored screw of the switch. Connect it to the dimmer's bronze-colored terminal (labeled *Common*). Connect the remaining two wires to the two brass-colored screws or pigtails. Connect all grounds.

2-Way, or Single-Pole, Dimmer

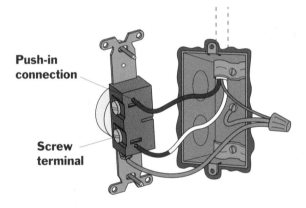

Push-in connection

Screw terminal

3-Way Dimmer

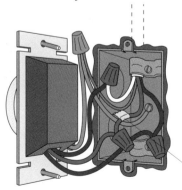

Mark "common" wire with tape.

TIP: Check that lamps are within the wattage rating of the dimmer. Don't use ordinary dimmers for fluorescent lamps—they require dimming ballasts and auxiliary controls.

Making Repairs

MAKING REPAIRS

Replace socket

Unplug the lamp. Loosen the socket's setscrew and pull out the entire unit, the cord attached. Separate it into the four parts illustrated. Disconnect its wires, identifying the wire connected to the silver-colored screw with a tape tag. Disassemble the new socket assembly. Unknot the cord, replace the socket cap, knot the cord, and connect the wires. Pull the cord down into the socket cap as far as it will go. Slip on the insulating sleeve, followed by the outer shell. Push the shell and cap together firmly.

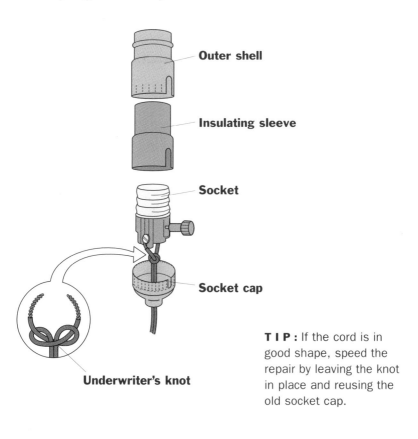

TIP: If the cord is in good shape, speed the repair by leaving the knot in place and reusing the old socket cap.

MAKING REPAIRS

2 Rewire lamp

Cut 6 feet of new cord, or enough to reach from the lamp socket to the receptacle plus 6 inches. Use the old cord to pull the new cord into the lamp. First cut the plug off the old cord. Split the ends of both cords about 6 inches and strip 3 inches from one wire on each cord. Cut off the other wire at the point of separation. Twist the stripped ends together and tape. Pull the old cord until the new cord extends about 12 inches beyond the lamp body. Separate the two cords and connect the new cord to the socket (see page 53).

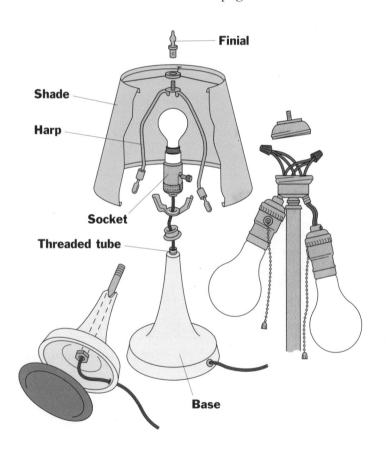

MAKING REPAIRS

Replace lamp plug

3

Buy the correct size of polarized plug. Remove its insulating cap and slip the plug body over the cord. Separate 6 inches of cord wire. Tie a knot where the wires join, and pull the cord back into the plug body. Mark the point where each wire meets its screw. Strip each at this mark and twist the wire strands together. Pull the cord completely back into the plug, loosen the screws, wrap the twisted wires clockwise 360 degrees around them, and retighten. Cut off the excess bare wire. Replace the insulating cap.

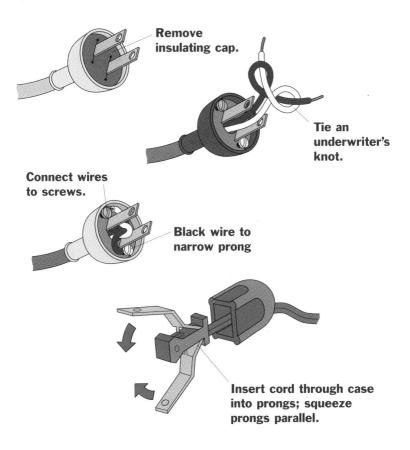

MAKING REPAIRS

Add cord switch

Size the switch to the lamp cord. Decide where it should be positioned on the cord. Make a small slit between the cord's two wires at this point, cutting only through the insulation. From the plug, trace the narrow prong's wire (the unribbed side of the cord) back to this slit. Cut completely through this wire at this point. Do not strip it. Open up the switch and lay the cord flat in the bottom (hollow) half so the cut wire's ends have at least an ⅛-inch gap between them. Squeeze the two halves together and screw closed.

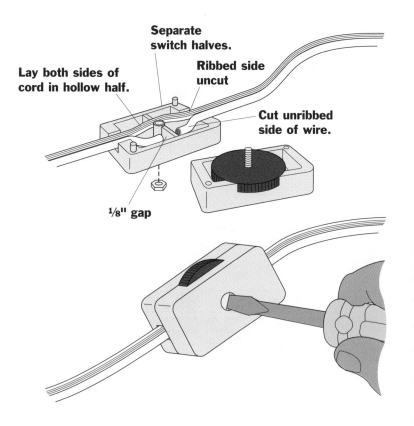

MAKING REPAIRS

Repair doorbell

Begin by tightening the connections to the button. If bell doesn't ring, unscrew the wires and touch their ends together. If the bell rings, replace the button. If there is no ring, check wires for frays or breaks. Then test the transformer with a voltmeter. If it measures no voltage, turn off power and replace the transformer. If voltage registers, check the screw terminals at the chimes. If voltage is reaching these terminals, the chimes are broken and need replacement. If not, the wires need replacing.

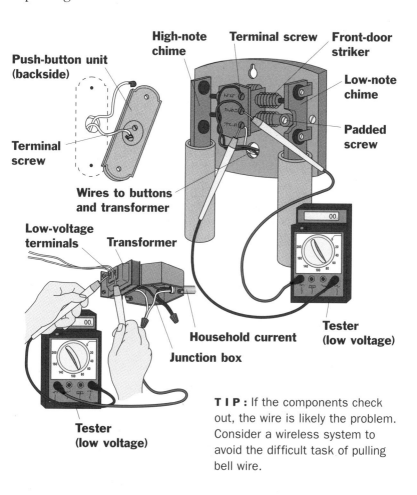

TIP: If the components check out, the wire is likely the problem. Consider a wireless system to avoid the difficult task of pulling bell wire.

57

Adding Improvements

ADDING IMPROVEMENTS

1 Wire smoke detector

For protection against alarm failure, replace a battery-powered smoke detector with a hard-wired 120-volt unit with battery backup. Install it in a ceiling-fixture outlet (see pages 19 and 23). Connect the same-colored wires and mount the detector directly to the box. Three-wire smoke detectors can operate in parallel so that all the units sound when one unit detects smoke. The black and white wires power the detectors, and the red wire connects the units' signal wires. The detectors must be the same make and model and on the same circuit.

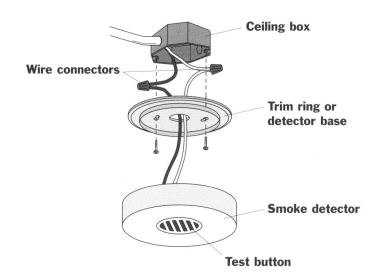

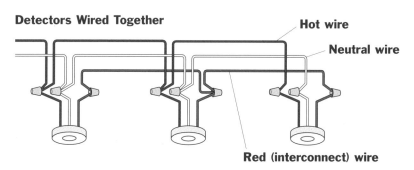

ADDING IMPROVEMENTS

2 Wire CO detector system

Carbon monoxide detectors monitor a home's air for the presence of dangerous levels of this odorless, colorless, lethal gas. A battery-powered detector mounted near a gas furnace and water heater offers protection at the source of a potential problem. More complete protection comes by wiring, in parallel, several 120-volt detectors with battery backup. Installation is similar to that for smoke detectors (see page 59). Follow manufacturer's instructions. The detectors must be the same make and model and on the same circuit.

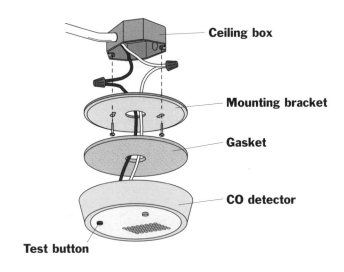

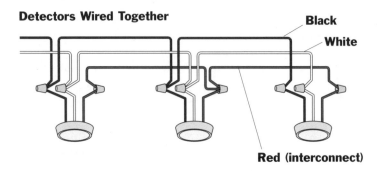

ADDING IMPROVEMENTS

3 Add motion detector

If possible, replace an existing switched light with the detector. Follow manufacturer's directions for installing the unit. Remove the existing fixture. Splice the original fixture's wires to the detector's wires, white to white, black to black. If not using an existing fixture, install a fixture outlet, following the instructions on pages 18 and 22. To complete the installation, screw the cover plate in place. Use a silicone caulk to seal the plate and mounting surface. Loosen the locknuts, adjust the lamp positions as necessary, and tighten. Screw in the floodlights.

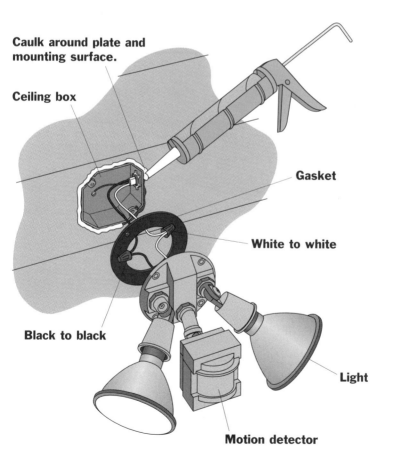

ADDING IMPROVEMENTS

4 Add surge suppressors

Surge suppressors protect electronic equipment from power surges. They're available as plug-in power strips and as surge suppression receptacles. With four to six outlets, power strips protect multiple pieces of equipment clustered in one room, but often create unsightly tangles of cord. Surge-suppressing receptacles are an attractive alternative when only one or two pieces of equipment need protection. Replacing a receptacle with the surge-suppressing receptacle involves a simple wire-for-wire installation.

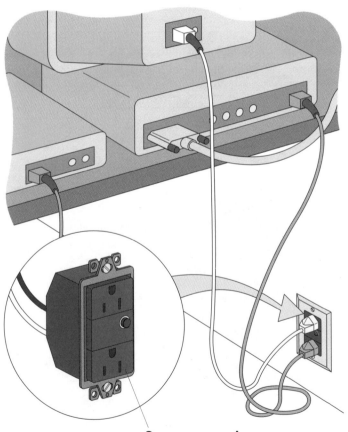

Surge suppression receptacle

ADDING IMPROVEMENTS

Add motor watchman

5

A manual, fractional starter protects motors against failure during power brownouts. If an extended overload occurs, the device's thermal unit, sized to the motor's rated current, automatically opens the switch, shutting off the power so the motor doesn't burn up. When this happens, the switch handle moves to the *off* position. The motor can't be turned on until it is pushed into the *reset* position. The device replaces the motor's regular switch. Installation or replacement is the same as for an ordinary single-pole switch (see page 27).

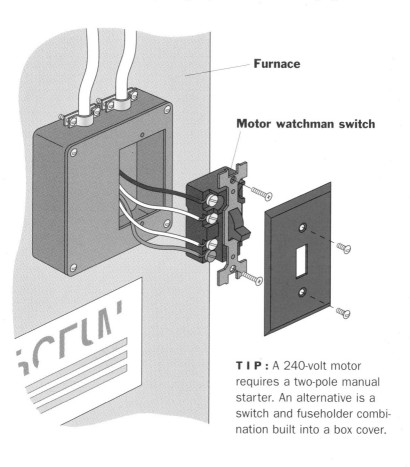

TIP: A 240-volt motor requires a two-pole manual starter. An alternative is a switch and fuseholder combination built into a box cover.

U.S./Metric Measure Conversions

Formulas for Exact Measures

	Symbol	When you know:	Multiply by:	To find:
Mass (Weight)	oz	ounces	28.35	grams
	lb	pounds	0.45	kilograms
	g	grams	0.035	ounces
	kg	kilograms	2.2	pounds
Volume	pt	pints	0.47	liters
	qt	quarts	0.95	liters
	gal	gallons	3.785	liters
	ml	milliliters	0.034	fluid ounces
Length	in	inches	2.54	centimeters
	ft	feet	30.48	centimeters
	yd	yards	0.9144	meters
	mi	miles	1.609	kilometers
	km	kilometers	0.621	miles
	m	meters	1.094	yards
	cm	centimeters	0.39	inches
Temperature	°F	Fahrenheit	5/9 (after subtracting 32)	Celsius
	°C	Celsius	9/5 (then add 32)	Fahrenheit
Area	in^2	square inches	6.452	square centimeters
	ft^2	square feet	929.0	square centimeters
	yd^2	square yards	8361.0	square centimeters
	a	acres	0.4047	hectares

Rounded Measures for Quick Reference

1 oz	=	30 g
4 oz	=	115 g
8 oz	=	225 g
16 oz = 1 lb	=	450 g
32 oz = 2 lb	=	900 g
36 oz = 2¼ lb	=	1000 g (1 kg)
1 c	=	250 ml
2 c (1 pt)	=	500 ml
4 c (1 qt)	=	1 liter
4 qt (1 gal)	=	3¾ liter
⅜ in	=	1.0 cm
1 in	=	2.5 cm
2 in	=	5.0 cm
2½ in	=	6.5 cm
12 in (1 ft)	=	30.0 cm
1 yd	=	90.0 cm
100 ft	=	30.0 m
1 mi	=	1.6 km
32° F	=	0° C
212° F	=	100° C
1 in^2	=	6.5 cm^2
1 ft^2	=	930 cm^2
1 yd^2	=	8360 cm^2
1 a	=	4050 m^2